What Every Kid Needs To Know About FIFO

A-Z Edition

Written by
Vicky Pellowe
The FIFO Family project

Illustrated by
Mirjana Segan

The FIFO Family Project acknowledges the Traditional Custodians of the lands and waters across Australia and pay our respects to Elders past and present. We recognise the strength and knowledge of Aboriginal and Torres Strait Islander peoples and honour their ongoing connection to Country.

A heartfelt thank you to the subject matter experts who generously shared their time and knowledge.

- Fiona Sands for her environmental expertise.
- Justin Forrest for his guidance on all things offshore.
- Peter Bellottie for sharing his Indigenous knowledge.
- And a very big thank you to Ross Jennings for his support across safety,

mining and many illustration reviews. This book would not be the same without your help.

Thank you to my family for their patience and creative ideas.

Another big thank you to the entire FIFO family community, for inspiring this book every step of the way.

We also acknowledge the support of the Government of Western Australia Department of Communities, whose funding helped bring this book to life. We are forever grateful for your support.

Government of **Western Australia**
Department of **Communities**

Copyright
ISBN 978-1-7640265-1-2

© The FIFO Family Project. All rights reserved.

No part of this publication may be reproduced, stored or shared without prior written permission from the author, except as allowed under the Copyright Act 1968 (Australia).

Dedicated to every FIFO worker, FIFO family & FIFO kid

Have you ever wondered what FIFO means?
FIFO stands for Fly-In-Fly-Out!

Some people have exciting jobs that take them far away for work.
They fly to job sites in places where there are no towns
and after a few days or weeks, they fly back home again.

But FIFO isn't just about working!
It's about adventures, friendships and families
finding special ways to stay connected.

In this book, you'll discover big trucks, cool careers
and even floating cities in the middle of the ocean!
You'll meet all kinds of FIFO workers and see how
families make the most of their time together.

So buckle up – your FIFO adventure starts now!

A is for Airports & Adventures

FIFO (Fly-In Fly-Out) workers help make mining, energy and resource projects happen in remote locations. Airports become like second homes for FIFO workers, with flights taking them to exciting workplaces across Australia and even around the world!

B is for BIG Machinery

From gigantic trucks to towering cranes, FIFO workers use incredible machines that make big jobs possible. Mining uses some of the biggest machines on Earth! From dump trucks as tall as a house to diggers that scoop up tonnes of dirt. Who knows - maybe one day you will drive machinery like this or even invent your own big machine!

C is for Community Connections

FIFO workers build strong friendships while working away, creating a community at work and building close bonds. On-site, they support each other through long days and tough jobs. Back home, it's just as important for FIFO families to connect with their local communities too. Together, everyone helps make the FIFO journey a little easier and a lot more fun!

D is for Different Routines

FIFO families have unique routines to make their lifestyle work – every family finds a rhythm that works best for them! Many use countdown calendars, marble counting jars or special activities, to mark the days until their loved one comes home. These routines help make the time apart go faster and the time together even more special.

E is for Environment Matters

FIFO workers and mining companies are always finding smarter ways to use resources, like recycling water, rehabilitating areas, planting trees and even funding research, which has led to whole new animal species being identified and protected – how cool is that?!

F is for FaceTime!

Staying connected with loved ones is a big part of FIFO life. Video calls, like FaceTime or Zoom, help workers see their family and friends, while they're away at work. Sharing pictures, texts or pre-recorded messages, is a fun way to stay connected and share special moments, even when you're not in the same place together.

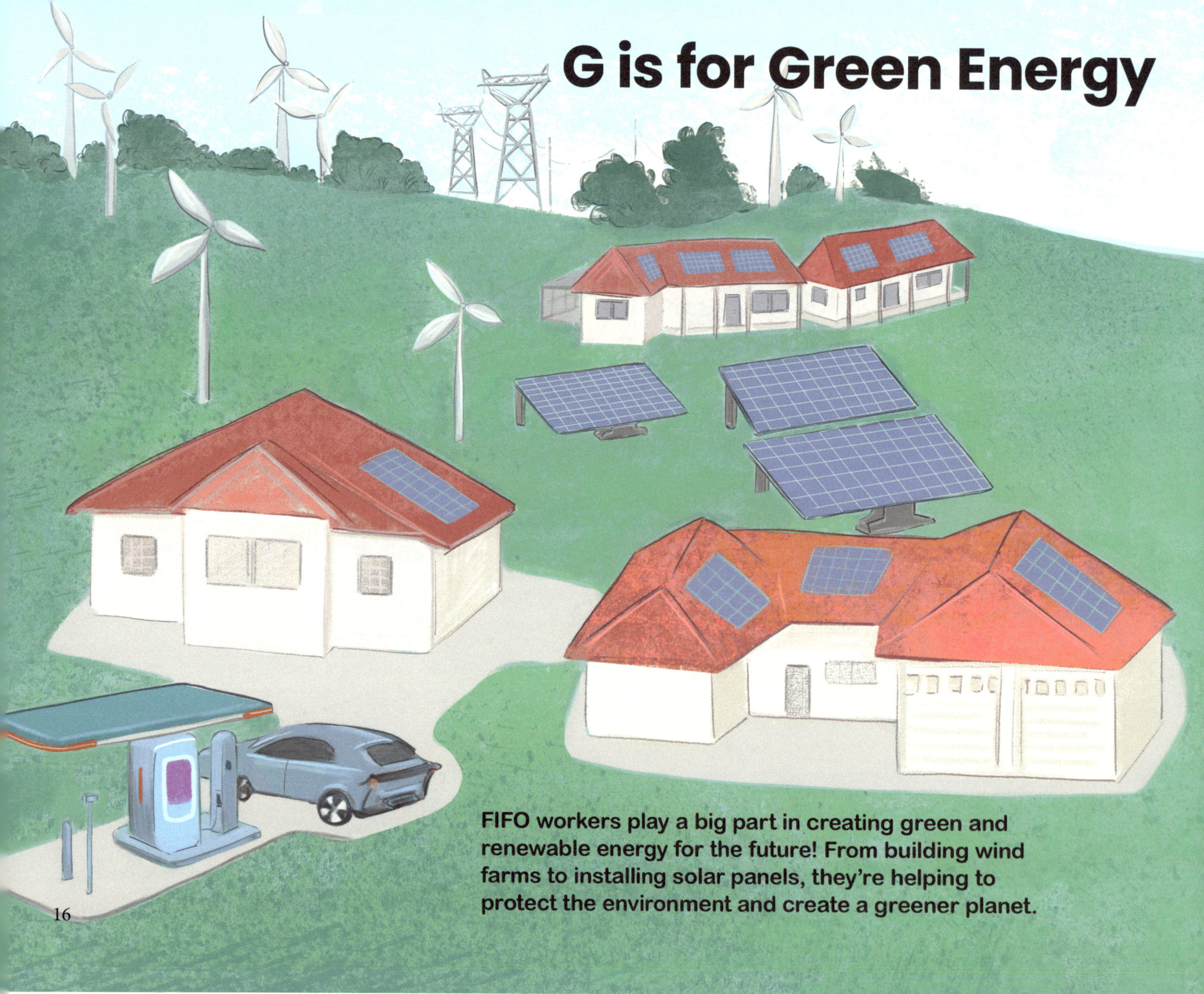

H is for Helicopter Rides

Can you imagine catching a helicopter to work? And living in the middle of the ocean for weeks on end?! Well, that's how offshore FIFO workers get to work! Offshore workers do important jobs, turning the ocean's resources into energy.

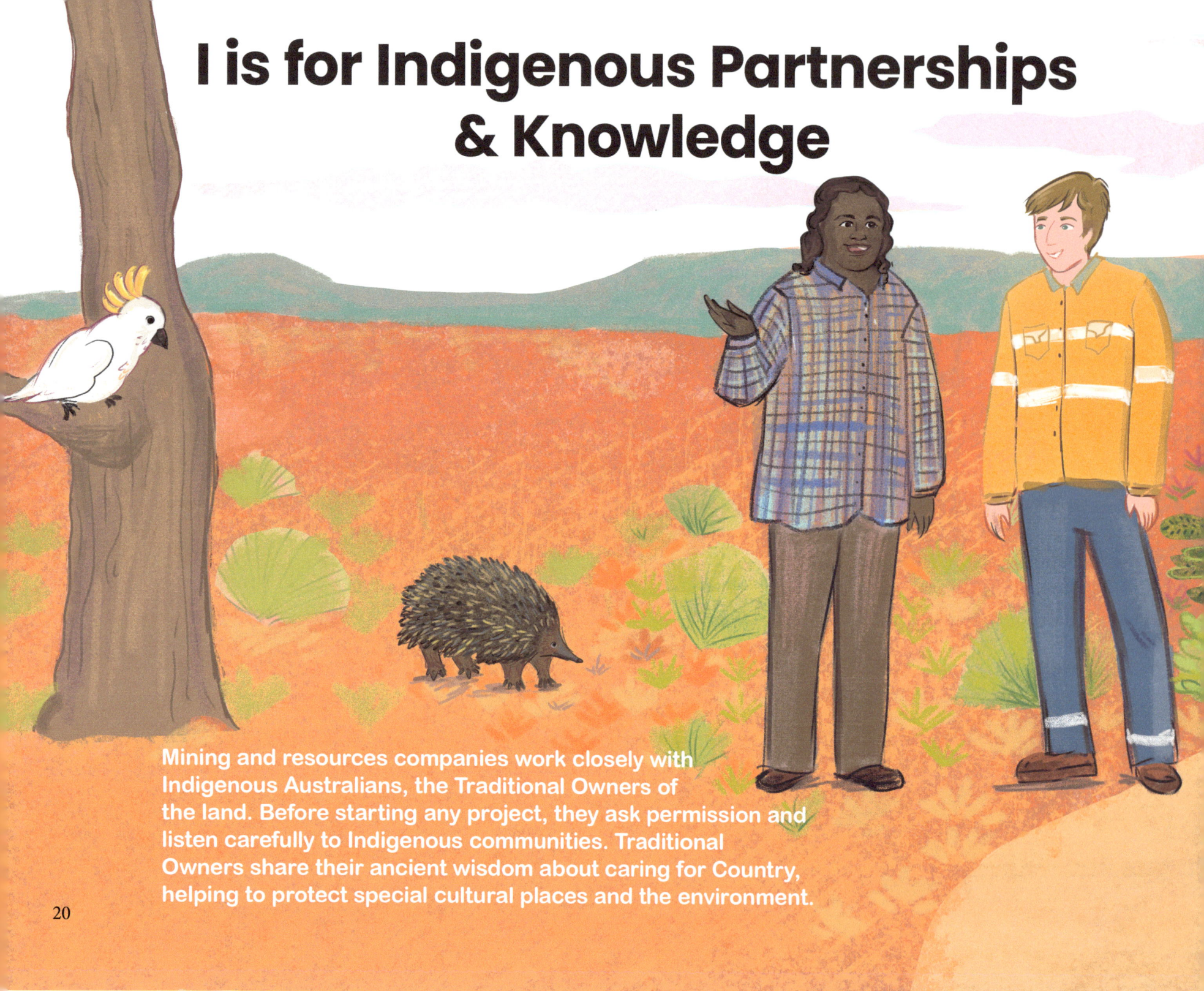

I is for Indigenous Partnerships & Knowledge

Mining and resources companies work closely with Indigenous Australians, the Traditional Owners of the land. Before starting any project, they ask permission and listen carefully to Indigenous communities. Traditional Owners share their ancient wisdom about caring for Country, helping to protect special cultural places and the environment.

J is for Jobs Galore!

There are so many different FIFO jobs like chefs, nurses, electricians, miners, administrators, engineers, truck drivers and even scientists. To name just a few! What would you like to do when you grow up?

K is for Keeping Safe!

Safety is the most important part of FIFO life. Workers wear helmets, boots and high-vis clothing and teams work together to stay safe. Everyone has a role to play - whether it's spotting hazards, doing emergency drills or checking gear. We all look out for our crew and workmates, to help make sure everyone gets home safely.

L is for Lights Out – Time for Night Shift!

Operations don't stop when the sun goes down… that's when the night shift starts! FIFO operations run 24hrs a day, 7 days a week, 365 days a year – even on Christmas Day! So that means many FIFO workers sleep all day and work their shift at night. As you can imagine, this can be quite tricky to readjust. So when they come home from night shift, it's important to get proper rest to reset those body clocks.

Hi, I'm Shannon! I'm a Mechanical Fitter, which means I fix and maintain the big machines that keep the resources sector moving. Sometimes I work during the day and sometimes at night - but no matter the shift, I'm part of the team keeping the site operating efficiently!

N is for Natural Resources

Australia has many amazing natural resources including gold, iron, copper and coal.
These materials help us build cities, power homes and create new technology.
Your phone, bike and even your toothpaste all come from mined minerals.
Mining plays a large role in all our lives!

"Hi, I'm Glen! I'm a FIFO Business and Contracts Manager and I help make sure everything on site runs smoothly. I work with lots of different teams, to make sure we have the people, supplies and plans we need to get the job done safely, on time and on budget!"

Gold
This makes your phone work with tiny wires, protects spacecraft from the sun and can be made into all types of jewellery.

Iron Ore
This gets turned into super strong steel to build cars, bridges, bicycles and even your house!

Diamonds & Gemstones
Used for beautiful jewellery making, cutting tools and even scientific instruments because of their durability and high strength.

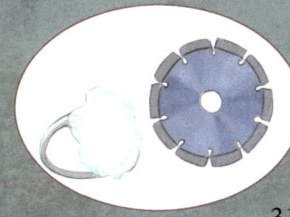

O is for Offshore Oil & Gas Platforms

Far out in the ocean, huge offshore platforms float above the waves like mini cities! Workers live and work here for weeks at a time, drilling deep under the seabed to find oil and gas formed millions of years ago. They follow strict safety and environmental rules to protect the ocean and all the amazing sea creatures they see - like dolphins, turtles and even whales!

P is for Problem Solvers

Do you like solving puzzles or building things? FIFO workers solve real-world problems every day – like how to dig safer, build faster or use less water and energy. With smart thinking and new technology, they find creative solutions to make mining better for people and the planet. Could you be a problem solver too?

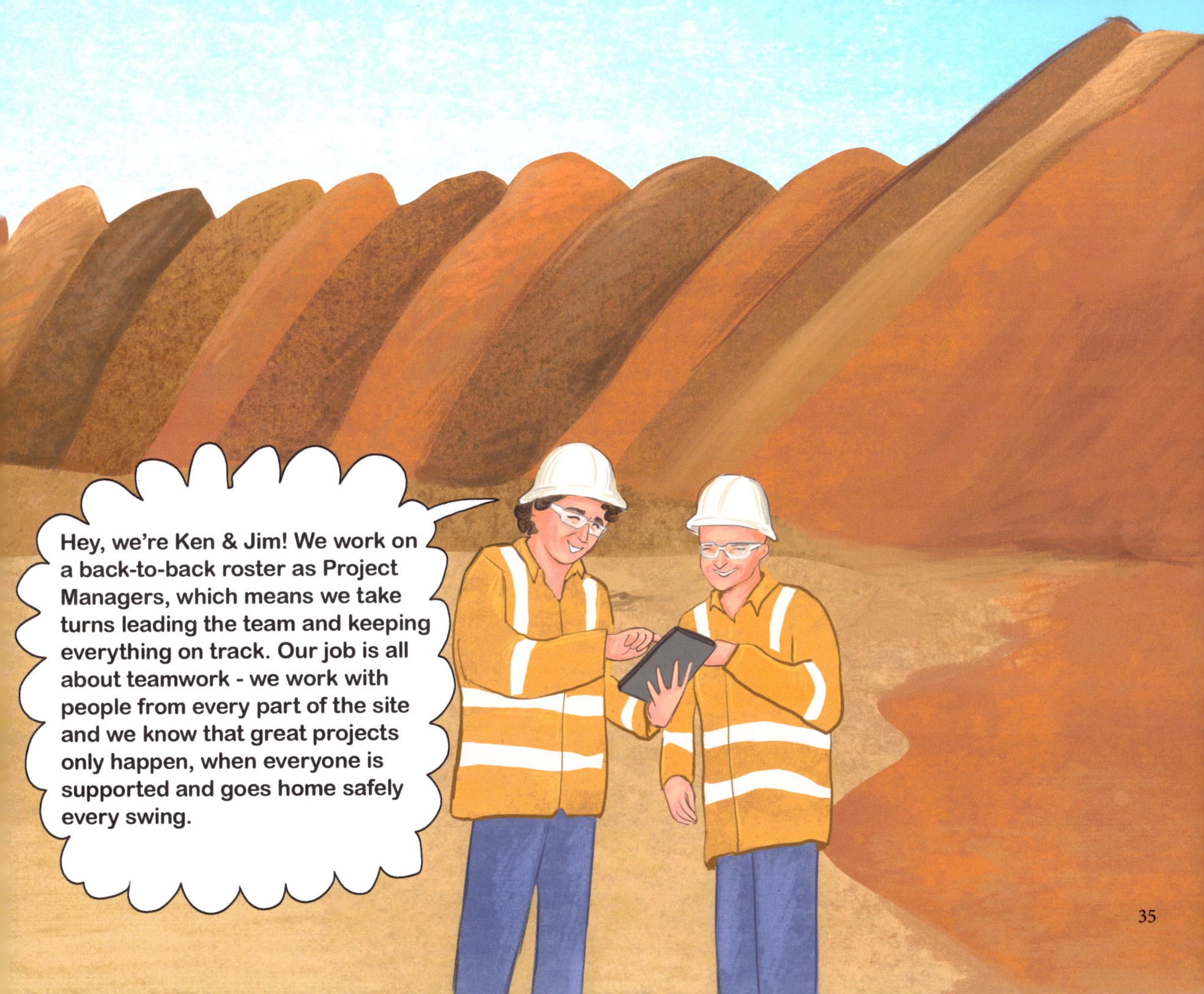

Q is for Quality Time

FIFO families make the most of their time together. It's not about how many days someone is working away - it's about the fun you have during R&R - like playing games, cooking meals or exploring new places. Every break is a chance to reconnect, recharge and enjoy quality time together at home.

R is for Rosters, Routines & Rituals

FIFO rosters tell us when workers are away and when they'll be home. Every family's roster and routine is a little different! Routines help keep things running smoothly and fun rituals – like 'Fish & Chip Fridays' or countdown calendars – make the time apart feel a little easier. From morning texts to end-of-day calls, it's the small things that help families stay connected.

Time for Fish & Chips

S is for STEM Superstars

Do you love solving problems, doing experiments or building cool things? Then FIFO might be the perfect career for you! STEM – short for Science, Technology, Engineering and Maths – is behind everything we do, from coding robots to testing soil and designing giant machines.

T is for Technology

Technology helps FIFO workers do amazing things – from giant machines that drive themselves to drones that fly over massive mine sites! Engineers, scientists and tech experts use smart tools like virtual reality, sensors and robotics to work safely, quickly and more efficiently. FIFO isn't just hard hats and boots – it's laptops, coding and big ideas in action!

U is for Underground or Open-cut

Mines come in two main types – open-cut and underground. Open-cut mines look like giant bowls carved into the earth and underground mines are like secret tunnels that go deep below the surface. Both use big machines and careful teamwork to safely find valuable minerals!

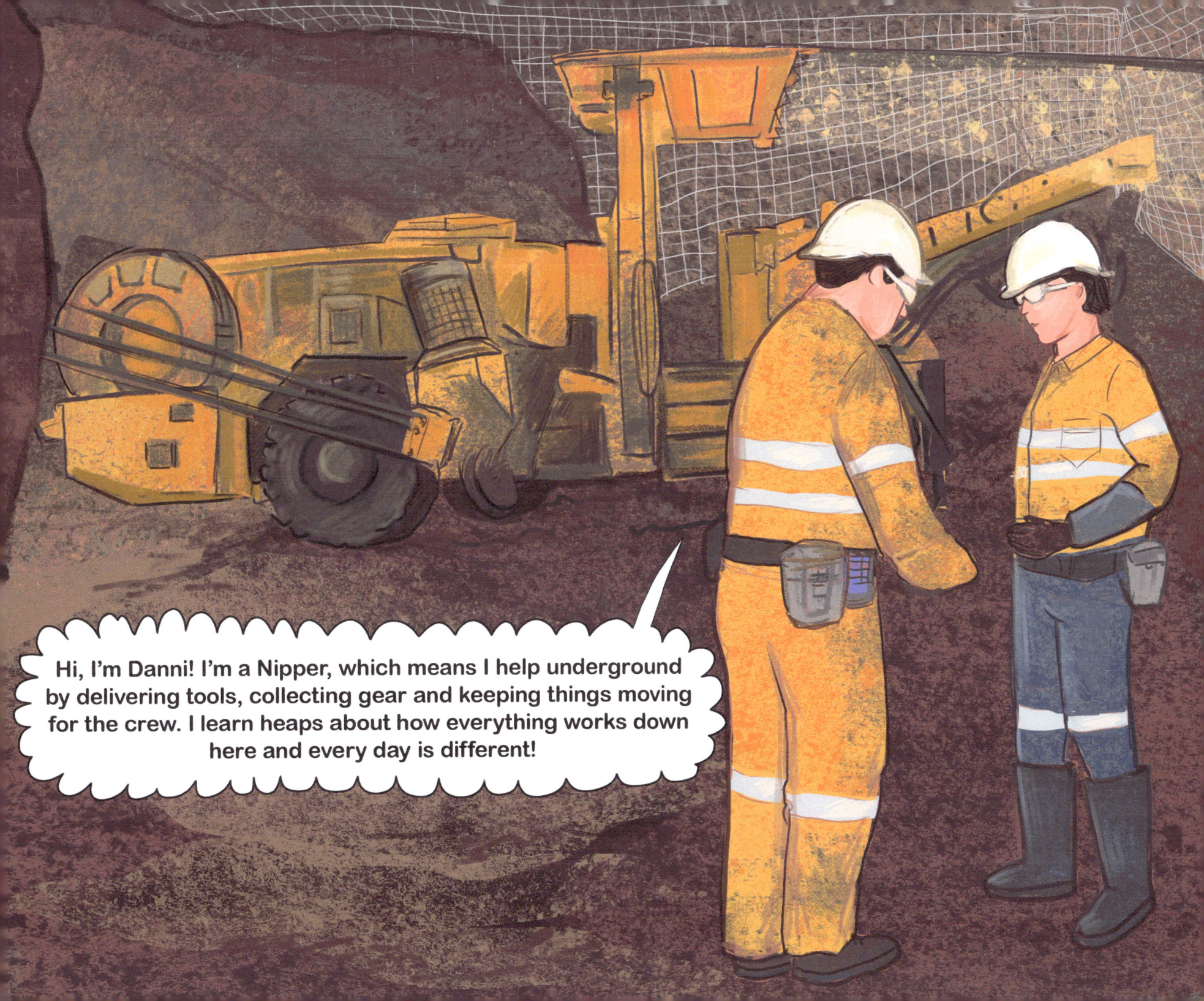

V is for Village

It takes a village to keep FIFO life running – on-site and at home. FIFO workers often stay in remote accommodation camps called "Villages", where everything from food to fitness is taken care of. But the village isn't just where you sleep – it's also the people who support you, from teammates on site to friends and family back home.

> Hi, I'm Hine! I'm a Utilities All Rounder, which means I help keep the village running smoothly for everyone on site. One day I might be cleaning rooms, the next I'm helping in the kitchen or making sure the laundry's done. We keep things tidy, comfy and ready for the crew when they come back after a long shift. It takes a team to run the village - and I love being part of it!

W is for Wild Weather & Wildlife

FIFO workers often work in interesting and remote places, where they might see wildlife like kangaroos, dolphins or sea turtles...and yucky ones too: FLIES! One day they might spot local lizards and the next they could face a heatwave, a dust storm or even a cyclone! Whether it's beautiful or extreme, nature is a big part of life on site, so FIFO workers need to be ready for anything.

X is for eXploration

Every big mining project begins with exploration! Surveyors help by carefully measuring and mapping the land, making sure geologists and engineers know exactly where to search and build. It's a bit like drawing a giant treasure map, where accuracy and clues lead the way!

Y is for Young Minds

The future of FIFO starts with young minds like yours! Whether it's inventing new machines, protecting the environment or solving big challenges – there's a FIFO job waiting for every curious kid.
What could your future look like?

Z is for Zero Limits

In the world of FIFO, there are zero limits to what you can dream, learn and achieve. Whether you want to build mighty machines, discover new resources or protect the planet – we need your ideas, energy and passion. The future of FIFO is wide open and it starts with you!

www.ingramcontent.com/pod-product-compliance
Lightning Source LLC
Chambersburg PA
CBHW041644220426

43661CB00018B/1290